James Ramsey Gibson Maitland

On Stocking Rivers, Streams, Lakes, Ponds and Reservoirs with

Salmonidae

James Ramsey Gibson Maitland

On Stocking Rivers, Streams, Lakes, Ponds and Reservoirs with Salmonidae

ISBN/EAN: 9783743398276

Manufactured in Europe, USA, Canada, Australia, Japa

Cover: Foto ©berggeist007 / pixelio.de

Manufactured and distributed by brebook publishing software
(www.brebook.com)

James Ramsey Gibson Maitland

On Stocking Rivers, Streams, Lakes, Ponds and Reservoirs with

Salmonidae

ON STOCKING RIVERS

STREAMS, LAKES, PONDS

AND RESERVOIRS

WITH SALMONIDÆ

BY

SIR JAMES MAITLAND, BART.

F.L.S., F.G.S., F.Z.S.

FOURTH EDITION

Published by J. R. GUY,

SECRETARY, HOWIETOUN FISHERY,

STIRLING, N.B.

1892

CONTENTS

PAMPHLET

ON

STOCKING

FOURTH EDITION

HOWIETOUN, *January 31st*, 1892.

THE last edition of the pamphlet on Stocking has been out of print for some months, while this issue has been delayed through the pressure of other work, and the necessity of practically re-writing the pamphlet to bring it up to date.

The first edition had to beg the whole question of modern fish-culture, the fourth edition finds artificial Stocking very generally adopted, and trout farms established in many places in England and Scotland as purely commercial ventures, thriving both fishculturally and financially.

The public, at least the angling portion thereof, is already very much up to date on the subject, the size and colour of eggs, the length and shape of yearlings or two-year-olds, are questions on which it holds a very decided opinion, nor is it any longer necessary to point out, as in the

first edition of this pamphlet, the gain to be
derived from stocking neglected pieces of water.
The public wants to know what each pond,
lake, or mile of stream should yield to the rod
annually, and what species, kind, and description
of fish will give the best sport.

And the public is right, because when once
the capacity—I might say the trout-bearing
capacity—of any given piece of water is known,
the trouble and cost of cultivation can easily
be adjusted to produce the desired results,
whether for profit or sport. I use the word
cultivation advisedly, because many still believe
that they have only to purchase a large number
of two-year-olds, and that then everything
necessary has been done. Alas! from a trout-
farmer's point of view, there never was a
greater mistake. Of course, in the case of a
new reservoir, like the Lake Vyrnwy, the lately
completed reservoir for Liverpool, it is only
necessary to plant a sufficient *number* of two-
year-olds, or their equivalent in yearlings or in
fry, and the desired result is produced ; but
in the case of ponds which have existed from
time immemorial, and in the case of streams, the
pre-existing conditions control not only the size of
trout required for the Stocking, but the mode of
cultivation to be followed to enforce successful
results.

To enable the reader more easily to under-
stand the practical rules which govern the successful cultivation of trout in ponds, lakes, and
streams, it will be convenient first to consider the
the *cost, manner of production,* and *uses* of Ova,
fry, yearlings, and two-year-olds ; next in order
the various conditions under which, each can be
most profitably used ; and finally, how any given
piece of water can be cultivated so as to yield
conditions favourable to one or other mode of
Stocking, and the management necessary to command the best possible results.

O V A.

*Its Production, Incubation, Appropriate Employ-
ment, and Cost.*

THE PRODUCTION OF OVA.—Most anglers would
ask, ' What can be said of the production of Ova ?
why not net the spawning grounds ?'   But a
very great deal can be written on the production
of Ova, and much that is new.   It is all very well
to net the spawning grounds.   And if you know
of several brace of magnificent spawners, and the
owner of the water will permit these being netted
and stript, well and good, the Ova are most
excellent, and if half the fry resulting therefrom
are returned to the water, well and good, both

owner and fish-culturist benefit, although when the cost of men's lunches, and nets, railway fares, and blank days are totaled up, the eggs seldom cost much under twenty shillings per thousand. The production of Ova, properly speaking, requires as much skill as the creation of a herd of pedigree cattle. Some good work may result in favoured waters from spawn promiscuously obtained; but the best results can only be ensured by the most careful adjustment between the class of spawners from which the Ova is obtained, and the nature of the stream, pond, or lake the produce of the eggs are intended to stock. To place the fry from eggs counting 25,000 to the gallon into high mountain tarns or rocky rivers is to court failure—there is not food enough in such places *all the year round*, to sustain such fast growing trout. Nor will much better results follow the planting of the produce of wild fish spawned promiscuously as they come to net in sheltered lakes or deep slowly flowing streams, where food is abundant, and fat trout or coarse fish numerous. The slow growing produce of small wild fish are, from their less size, subject to the cannibalistic attentions of the existing older stock for many months longer than the quick growing produce of wild breeders which have been carefully selected, or the produce of pedigree trout, whose age and parentage are known with scientific accuracy.

The produce of breeders when confined are looked on with distrust especially in Germany ; and we think with much reason, *if* and *where* the breeders are *wild fish* caught and *confined* for spawning purposes, for the same reason that wild animals in zoological gardens are unsatisfactory as breeders.

But when the trout are domesticated, perhaps for four or five generations, when their characters are studied, and when individual trout are as well known as prize rams, or a Field trial setter—and their parentage not one whit more uncertain—when, in fact, they are not only domesticated, but bred to points in the same manner as a Clydesdale or a short-horn, the fish-culturist gains the power to produce eggs of varying qualities, and selects those best suited to whatever water he desires to stock. The confinement of the breeders ceases to be a makeshift, and becomes a primary factor in improving and adapting the race of trout to the high standard of modern necessities and modern requirements.

The production of Ova thus comes to depend on the establishment of a trout farm, farmed on scientific principles, and containing ponds large enough and deep enough for the heaviest spawners. The ponds must be numerous, so that the various ages may be separated, and it is advantageous even to separate the sexes of the older trout. Males do not fight if there are no spawners to fight about, and the ugly wounds, so prevalent in autumn and so fertile a source of disease, are conspicuous by their absence.

Sufficient ponds having been provided, the strain of trout requires to be obtained from selected breeders, selected yearlings, selected two-year-olds, and again selected three-year-olds. After this fourth elimination of the less fit, the trout may be allowed to grow for two seasons, after which they are fit for spawning and the Ova are of good size; but the best and almost the largest Ova are yielded in the seventh season. The Ova in the eighth are very slightly larger, but the impregnation is not so perfect, and the number of Ova per pound of trout is less. The next season the Ova are still very good, but the number per pound is sensibly smaller, and percentage of impregnation perceptibly lower. We believe the lowering of the percentage of impregnation in very old trout is due to the thickening of the shell of the Ovum.

It is necessary to have a series of ponds for each strain of trout, although so many ponds are not required for ordinary strains as for the selected pedigree breeders, nor is it necessary to size the breeders so frequently, but it is a mistake to suppose that the same class of ponds are suited to all strains of trout. Some prefer the water to flow in as a stream ; others thrive better if the water falls into the pond over a perpendicular inlet. *Char* live in deep-shaded ponds, narrow and constructed east and west, so that the south side gives shade at all times, while they die only too freely if the pond runs north and south, and is difficient in shade. We have found *S. fontinalis* thrive, however, in a deep, narrow pond, with perpendicular sides running north and south, this pond is well shaded by overhanging trees ; while in two similar ponds, not so shaded, the loss both in spring and after spawning is very heavy.

The production of Ova also includes the introduction of fresh blood and the production of fertile crosses. There are now at Howietoun many thousand yearlings whose pedigree shows one-eighth salmon blood and seven-eighths trout blood. And we have reason to believe that this cross will prove perfectly fertile *inter se*—although the higher proportions of salmon blood have been very largely infertile ; but this subject will be more fully recorded and discussed in the second part of the history of Howietoun.

THE INCUBATION OF OVA.—The strength of the fry is controlled by the mode of incubation of the Ova. The best Ova from the best strain of trout may easily be so treated that the resulting fry are puny and ill-conditioned ne'er-do-weels. There are three secrets of incubation, these are :—no *strong* light ; no *sudden* change of temperature ; no *stagnant* water. Strong light injures the nervous system ; sudden change of temperature injures the veinous system ; stagnant water injures both. The light can be controlled with a little care ; the even temperature can be insured by incurring some expense. But to render the eggs secure against stagnant water requires both great care and much expense. The hatching trays must be so arranged that a current of water is constantly impinging on each individual ovum and sweeping off (the pores of the shell) the carbonic acid continually produced during the formation of the embryo, but more particularly in the few weeks immediately preceding hatching.

Many designers of hatching apparatus seem to have had in view merely how to obtain the greatest results on *paper*, the value of these results does not appear to have received much consideration. This or that apparatus will incubate, and perhaps actually hatch so many thousand eggs, and it requires little water and takes up next to no space ! The important question, what is to be done with the fry, never seems to have occurred to the designer. Now the fact is that the fry require nearly two square feet of surface per thousand if they are afterwards to be fed in the hatching troughs, and fully one square foot of surface if they are merely to be retained until the absorption of the yolk-sac. But this space is sufficient for double that number of eggs; therefore there can be no real inducement to design hatching trays to incubate an excessive number of eggs in the smallest possible space. However, the practice of over-crowding trout and salmon eggs is general, and to this practice many of the disappointments encountered in private hatcheries are due. If the eggs are so incubated that a current of water perpetually cleanses each egg during the whole process, strong fry is the result; if not, the contrary obtains.

APPROPRIATE EMPLOYMENT OF OVA AS A MEANS OF STOCKING.—Ova is now generally forwarded on the point of hatching, and for those who have troughs suitable for rearing alevins to the feeding stage, or fry for four or five weeks longer,—nothing can be better than fully eyed Ova laid on temporary trays, or on the bottom of the trough. Only if fully eyed Ova is laid on the bottom of the troughs, certain precautions are necessary, as the embryo is very easily suffocated —drowned in fact—in still water. These precautions are—first, to spread the eggs evenly so that no eggs lie on the top of others, or even touch others latterly; second, to lower the surface of the water to within an inch of the eggs; and thirdly, to arrange the inflowing supply in such a manner that the current cleanses the carbonic acid from every egg—the water being of course raised to three or four inches depth so soon as the eggs are well on the hatch. If the troughs are provided with trays or grilles, the eggs may be spread on them and thus do away with the necessity for lowering the water, but in any case the supply of water should be equal to three gallons per minute for each 10,000 eggs in any one trough, and this quantity may be safely passed through any number of troughs not exceeding five. But the most appropriate use of ova as a means of stocking is by sowing it down in redds. There are many lakes so remote from railways and roads that practically it is impossible to convey a sufficient stocking of yearling to them. In this case redds become a valuable adjunct. There are few lakes that have not some feeding stream suitable for naturally rearing fry—here again a few redds, proportioned to the extent of the rearing ground, should be used, thus ensuring the best possible results at a minimum of cost.

The construction of redds and the selection of suitable places will be found at page 54.

Cost of Ova.—The cost of Ova is composed of two items—one fixed—namely, the cost of incubation, and the other, very variable. The cost of in-bation is fixed for all trout and salmon Ova per gallon, and for each class of Ova per thousand ; but as the same space is required in practice for every gallon (at Howietoun this space is two boxes), while in different classes of Ova the number to the gallon varies from 23,000 to 50,000, it is evident that the space required per thousand, and the consequent cost of incubation per thousand is nearly double per thousand for eggs running 23,000 to the gallon to that for eggs running 50,000 to the gallon,—in practice the limits are first-class salmon, 23,000 ; first-class trout, 28,000 ; eggs from young salmon and grilse, 30,000 ; ordinary trout, 40,000 ; well-bred trout, 32,000 ; ordinary wild lake trout, 40,000 ; wild hill-burn trout, 50,000. Hill-burn trout can be improved to produce eggs running 40,000 to the gallon.

The principal hatching house at Howietoun cost a little over £3000, and as at present fitted incubates 80 gallons of Ova. The cost of renewals of hatching houses is about 14s. per annum, and the proportion of wages is about £2 per week for say 15 weeks, as in practice the house cannot be all filled under five weeks, and the eggs require about ten weeks for incubation. Moreover the house and boxes are used for fry for a further period of fifteen weeks. The calculation will stand thus :—

Interest on £3000 at 4 per cent., . .	£120	
Depreciation and renewal calculated on 80 gallons of Ova at 14s., . . .	56	
	£176	Half = £88.
Of which one-half is credited to use for fry leaving to debit of Ova, . . .	£88	
Add Wages—40s. for 15 weeks, . .	30	
	£118	

Or, dividing by 80 gallons, £1, 9s. 6d., that is, £1, 2s. and 7s. 6d. per gallon respectively.

A slight reduction may be made on these figures on the ground that a certain number of the boxes can be used twice ; but as this interferes to some extent with their being afterwards used for rearing fry I have neglected this item. Now, £1, 9s. 6d. per gallon is for good trout eggs running from 26,000 to 32,000 per gallon—almost exactly 1s. per thousand—and this may be taken as an accurate base in calculating cost. The cost of a smaller hatching house will probably be not less than the proportional £37, 10s. per gallon,[1] the labour, however, will be unduly increased, and, unless the house is built of substantial masonry, 4 per cent. on cost will be too little to debit as rent ; and in all small houses the cost of the water supply is so important an item, that unless carefully considered before the construction of the building is commenced, the cost of incubating

[1] The prime cost of the Hatching-house per gallon is £3000, divided by 80 gallons, or £37, 10s.

eggs is more likely to be over 1s. per thousand than under.

The second item is the cost of obtaining and fertilising the Ova. I have had large experience in obtaining salmon Ova, and in different districts the cost varies. On the Forth there are many blank days; on the Tweed few; and on the Tay practically none. I think it is safe to assume that the cost of collection of salmon ova may average for the season anything between 3s. and 10s. per thousand, while the 10s. figure closely approximates the cost of collection of wild trout Ova. Even at Loch Leven itself, in the earlier days of my fishcultural work, the whole Ova collected during the season, divided by the whole cost of collection, more than once gave a figure slightly exceeding the above estimate. Of course, now, through long experience, Ova is collected at the Loch at a much lower figure. An accurate estimate of the cost, when the breeders are confined, may be obtained by adding rent of ponds, labour, food, and value of stock together, and dividing by the number of eggs obtained.

At Howietoun, four ponds are devoted to pedigree *S. Levenensis* breeders,—ponds Nos. 9, 10, 11, 12. No. 9 pond has 840 sq. yds. of surface. Each of the ponds Nos. 10, 11, 12, have 1500 sq. yds., making 5340 sq. yds., or 1 acre, 0 roods, 16 rods, and 16 square yards of water surface; and each pond is from 8 to 14 feet in depth. These ponds are laid out for the following stocking :—

S. LEVENENSIS PEDIGREE BREEDERS.

Showing Position in Ponds as at 1st May.

Year.	Age of Spawners at 1st Dec.			Pond 9.	Pond 10.	Pond 11.	Pond 12.	Remarks.
1891,	7	5	3	*	89	87	85	No. 10 pond not spawned.
1892,	8	6	4	91	89	87	85	No. 9 stocked with pedigree yearlings.
1893,	9	7	5	91	89	87	85	
1894,	8	6	4	92	89	87	91	No. 9 stocked with selected pedigree two-year-olds.
1895,	7	5	4	94	89	92	91	No. 9 stocked with pedigree yearlings.
1896,	8	6	5	94	89	92	91	
1897,	7	6	4	95	94	92	91	No. 9 stocked with selected pedigree two-year-olds.
1898,	7	5	4	97	94	92	95	No. 9 stocked with pedigree yearlings.
1899,	8	6	5	97	94	92	95	
1900,	7	6	4	98	94	97	95	No. 9 stocked with selected pedigree two-year-olds.
1901,	7	5	4	1900	98	97	95	No. 9 stocked with pedigree yearlings.
1902,	8	6	5	00	98	97	95	
1903,	7	6	4	1	98	97	00	No. 9 stocked with selected pedigree two-year-olds.
1904,	7	5	4	3	98	1	00	No. 9 stocked with selected yearlings.
1905,	8	6	5	3	98	1	00	

Note.—The age of Spawners at 1st December is taken as age *next* birthday, that being within six weeks of actual age at spawning. With this exception, ages are always counted as from the 1st of January of the *year in which the fish commenced to feed*, whether hatched in the preceding month of December, or on or before the 1st of January of that year.

The preceding table at once forms the basis for an estimate of the cost of the production of trout eggs. The cost of the alternate stocking of pond No. 9 is £125 for 5000 two-year-olds, and £55 for 5500 selected yearlings, or £180 spread over three years (say 1894-5-6), making £60 per annum. The labour required would be at least two skilled men at £60; making £120 per annum, and about forty weeks of two women at 10s. per week, opening shell-fish to feed the trout, —for unless the trout are properly fed they do not yield good ova—that is, 80 weeks at 10s., equal to £40, making the whole labour bill £160 per annum. The food bill for this number of trout of the ages noted in the table would not be less than 40 tons of clams, their present cost at Howietoun including carriage is just under £5 per ton. Thus the food would average £200 per annum. The water supply for these four ponds should not be less than 2,000,000 gallons, and should average about 3,000,000 gallons per diem. It is difficult to place a value on this item, as so much depends on the first cost of the reservoir and inlet works. But taking this at one-tenth of a penny per thousand—3,000,000 gallons represents £456, 5s. per annum—and the water supply—which, for this purpose, it must be a supply of *pure* water—must always form one of the largest items of cost in the production of Ova.

The two items remaining are rent and management, if the cost of construction is considered part of the water supply, and is included in the moderate calculation of one-tenth of a penny per thousand gallons. The ponds, with the necessary ground between and around them, will require between two and three acres, and the interest of cost of roads, fences, and buildings, mincing machinery, etc., falls to be treated as rent. I do not think that less than £30 per annum can be calculated on, omitting management, risk, and profit, the balance will stand :—

			Per Annum.		
Stock,	.	.	.	£60	0	0
Food,	.	.	.	200	0	0
Labour,	.	.	.	160	0	0
Rent,	.	.	.	30	0	0
Water supply including construction of ponds,			.	456	5	0

£906 5 0 or say £900 0 0

The estimate yield of good Ova is 150 gallons per annum, which at £6 per gallon is   . £900 0 0

But if only 100 gallons are obtained, and without good management this is not improbable, the cost rises to £9 per gallon, since 100 gallons at £9 per gallon is, ... £900 0 0

If the annual cost of water and the four ponds are found by taking 4 per cent. on the cost of construction, and adding 10 per cent. for renewals, which, as the ponds require clean earth on the bottom and sides every fifth or sixth season, is not too high a percentage, the sum estimated will rather be increased than dimished.

When Ova are obtained from private ponds or lakes, my experience is that the cost of collection is much above the highest figure brought out above, viz., £9 per gallon. This, with the £1, 9s. 6d. per gallon,[1] brings the prime cost of eyed Ova to £10, 9s. 6d. per gallon, or at 30,000 to the gallon, 6s. 6d. approximately. When the risk, cost of packing, freight, and management are considered, I am of opinion that first class Ova will tend to improve in price. The present wholesale price, including delivery, viz., £37, 10s. per 100,000, being exactly 7s. 6d. per thousand.

I have given those estimates in the above form chiefly in answer to many inquires from members of District Fishery Boards, many of whom it was impossible to answer fully by letter.

FRY.

Its Production, Appropriate Uses, and Cost.

THE PRODUCTION OF FRY.—Salmon or trout fry may be produced by forming artificial spawning ground in connection with rivers, lakes, or ponds. In salmon districts there generally exist many natural spawning places inaccessible to salmon. The introduction of salmon to these spawning streams can usually be successfully accomplished by opening the upper waters through

[1] *See* page 18.

the removal of obstructions or construction of passes. And beyond all doubt in most cases a great public benefit would accrue. Be that as it may, in Scotland at least, many hundred miles of excellent spawning ground is annually wasted, as far as salmon are concerned. The power and the inducement to open up these spawning grounds is a matter for legislation, and in questions of salmon rights, legislation moves but slowly. The interference with existing rights, the absence of right of salmon fishing in the riparian owners of the as yet salmonless waters, and the difficulty of assessing for the heavy cost of opening these waters, all act as a drag. .

It is not within the purpose of this pamphlet to consider the legal aspect of the difficulty. But on the supposition that an existing obstacle was removed, and miles of new spawning ground opened up, the probable effect would be that the old fishery district, as a whole, would benefit out of all proportion to, the rod fishing created above the obstacle, and the owner of the fishery immediately below the obstacle might be presently injured; and probably, after a few years, substantially benefited by the increased number of fish in his water, due to the produce spawned in the new water returning for a like purpose.

Until one corporate body *in each district* has sole power over the net-fishing—so that one or more

tacksmen fish nets in saltwater, and only as sett[1]
by the said corporate body, net and cruive fishing
in fresh water being abolished, and rod fishing
for salmon vested in the said body under such
limitations as will prevent any gross interference
with either private or public rights—there is
little hope of any rapid improvement of salmon
fisheries by scientific fish-culture.

Each district can only produce a certain gross
weight of fish, and this can be more cheaply and
quite as certainly captured from half a dozen
stations as from a hundred. The total number of
fish caught depends on the gross produce of the
rivers in the district. The multiplication of fishing
stations is merely a needless multiplication of the
expense—not to increase the total catch, but to
decrease the value of rights already existing, and
in this matter of late years on the Scottish coast
the Crown has not been blameless. To those who
know the cost of wear and tear of sea nets, the
suggestion that the profits of the Scotch salmon
fisheries would be more than doubled under such
management will not appear unreasonable, and the
equalisation of the assessment that would then be
practical would enable fish-culturists to double or
treble the total yield of salmon in the district,
this yield being controlled by *the available area
of the feeding ground for salmon fry*—ground of

---

[1] This term refers to the conditions of the lease.

which fish-culturists, with the assistance of artificial redds, would take the fullest advantage.

On the production of fry in the hatchery, I
cannot do better than quote from my *History of
Howietoun*, Part I., p. 57, line 13 :

'The hatching-boxes are stripped of their fittings,
thoroughly cleaned, and the safety-screens fixed 13 inches
above the outlet. If the ova is on the point of hatching it
is then laid down on the wood. It is often convenient to
prepare the boxes a few days before the eggs are due to
hatch ; and since, if the eggs remained many hours in the
comparatively still water at the bottom, some of the embryos
would be drowned, a small piece of tile is laid in the box,
and a couple of grilles balanced on it until the hatch com-
mences, when the tile is taken out, and the grilles reversed
and removed. With 3 inches of water in the hatching-boxes
there is no danger of the alevins becoming suffocated—a very
common source of loss in badly constructed trays.

'A CO-OPERATIVE BREATHING SOCIETY.—Shortly after
hatching the alevins congregate together in dense masses
in the corners and against the sides of the hatching-boxes.
If the water over these masses be slightly discoloured so as to
render visible the paths of the currents, small whirlpools will
be noticed descending over the apex of each mass, while from
the base the water radiates a thin swiftly-moving cloud.

'On closer inspection it will be noticed the alevins are
lying with their heads approximately towards the apex, and
their paired pectoral fins working with remarkable rapidity ;
thus, instead of those lowest in the mass being in danger of
suffocation, they really receive the strongest water currents. The
mass is, in fact, a gigantic co-operative breathing association.

'Very different is the fate of the unfortunate who finds a
refuge beneath the sheet of glass so frequently used in the
early days of trout-hatching to preserve the eggs from fungus.
In a very few hours a pale lifeless form marks the grave of
the suicide. Charred wood has abolished the necessity for

linings of any sort, but the general public still delight in providing all sorts of death-traps, under the name of shelters.

'After the fry have herded together for a few weeks, more or less, according to the temperature of the water, they are seized with a roaming and inquisitive spirit. At this stage the use of flannel in fixing the safety-screen becomes apparent. But it must be new flannel; old sodden stuff is certainly water-tight, but still the fry manage to burrow beneath it, and, although they cannot pass through, manage to suffocate themselves in considerable numbers. New flannel is seldom water-tight, but this is of no consequence. Whether the fine hairs annoy them, or whether inherited instinct teaches them to dislike the manufactured product of the animal whose periodic washings have so frequently nauseated their ancestors, I know not; but it is nevertheless a fact that the alevins will test every joint in the box, but leave the flannel-protected joints of the safety-screen alone.

'The only attention newly hatched alevins require is as to the strength of the current passing through the safety-screen. If the water is kept shallow, and the current equivalent to two gallons per minute through the ordinary Howietoun hatching-box, there is some danger, during the first few hours of the hatch, of so many of the holes of the perforated zinc of the safety-screen becoming clogged with cast-off shells as to make the current sufficiently strong through the remainder to catch and hold any unfortunate alevin passing. It is not safe to heighten the water in the hatching-box by placing flannel on the outlet screen before nine-tenths of the eggs are hatched, or the decreased current will cause partial suffocation of the embryos. And even when one-tenth of the eggs are left, they should be well separated with a feather before the water is deepened. If alevins get against the safety-screen, the pressure of the water gradually forces the yolk-sac, which is very elastic, through one of the perforations of the zinc. If No. 9 size is used, little harm accrues, as a strong alevin, so soon as he feels the process commence, can free himself by his struggles; and he takes care in future to avoid the screen·

But with a smaller size matters are very different. The least portion of the yolk-sac slipping through—which it does the more easily, as the smaller the number of zinc the stronger the current—bulges out, and keys the fish on the reverse side of the safety-screen in such a manner that escape is impossible.

'At Howietoun the eggs are laid down to hatch between nine and ten in the morning, and unless the day is very cold, the hatch is completed by three in the afternoon, when the shells are skimmed off and the water in the box heightened. This is tedious work, and we do not care to hatch more than a quarter of a million in one day, although it is occasionally necessary to lay down a much larger number. After the shells are skimmed off and the water heightened in the boxes, the attention required is reduced to a minimum. As the fish get stronger the current is increased, the regulating tap being moved once a week through a space equivalent to one quart per minute per box. Hardly any alevins die except in experimental lots. The boxes themselves never require cleaning; the alevins are their own housemaids. They are constantly scouring over the bottom of the box, and keep the charred wood polished like a piece of dark mahogany. The current carries the dirt through the safety-screen, and it settles in the space of 13 inches between that and the outlet. The rough of this dirt is removed with a syphon every morning. Once a week the cork in the settling-tank is drawn, and the bottom of the box thoroughly cleaned with a brush.

'The importance of exactly balancing the number of alevins to the box, and adjusting the proportions of the hatching-box to the current, are apparent. There is little danger of having too many alevins in a box, as the number put in is determined by the number of feeding-fry that can be properly reared; but if too few alevins are placed in the box, or if they have been improperly incubated and have not sufficient vitality, they will neither have strength nor activity to polish the bottom; and as—with due deference to the opinions of some—perfect cleanliness at this stage is absolutely essential to their future success in life, if they do not do their own housemaids' work it must be done for them at great trouble and expense. Up to this stage the daily work in the hatching-house is much the same for eggs and alevins. It is not advisable to begin too early in dark mornings; eight o'clock is quite soon enough to unlock the door. The Manager should always enter the house first, check the thermometers, and notice the overflows of the regulating-tanks. The girls then look over the hatching-boxes, pick out the opaque ova, and note the numbers on the printed form. The Manager gives the head attendant a list of the boxes from which all unimpregnated eggs are to be picked, for sale or before laying down to hatch. If it is a bright morning he sees that the swung shutters are tightly closed on the south side of the house. Direct sunlight is not only injurious to the embryos, but is apt to induce a cryptogamic growth on the shells of the eggs. The particular fungus I have not determined, its principal characteristic being the length and delicacy of its filaments.

'When any eggs are near hatching, boxes which have been already emptied, for sale or otherwise, are prepared; if not, the Manager is free to go to the ponds. Should the following day be one on which ova are despatched, he returns in the afternoon to throw the eggs on to the frames; but on other days he merely looks round in the evening to receive the schedule of the dead ova picked out, and to see generally that all is right. When the alevins are hatched, he has to attend to the depth of water in each box, which is increased by raising the flannel on the outlet screen, and to the weekly increase of the supply when the water is raised. He has also to see that a sufficient number of ova packing-boxes are prepared, with sawdust carefully filled between the inner and outer cases. In practice, it is found necessary to have at least twelve ordinary egg-packing boxes and six foreign always ready in the box-room. He also requires to take stock of the quantity of swan's-down squares, and to check the amount of sphagnum moss in the cellar, where it keeps best.

'It must always be remembered that, in the case of a heavy

snow, it is impossible to get good moss sometimes for weeks together. If the stock runs short at these times, marshy places below springs are generally open, but the sphagnum is rank, soft, and the lower portion frequently bleached, and neither suitable for felting nor capable of living over a long sea voyage.

'The Manager inspects the grilles removed the day previous, to see if they are in good condition and have been properly cleansed, after which he superintends their being placed on the rafters of a shed, where they remain dry and safe all summer.

' When the fry begin to feed, the hatchery demands much more of the Manager's time. He requires to check the food left by the butcher, test the paste prepared for feeding the fry, and to specify the exact quantity of food each box requires.

' FEEDING THE FRY.—The best and most economical food for trout fry costs about 1s. 4d. per lb., and, strange though it may appear, it is much cheaper than liver at 1d. per lb.,—that is to say, one pound of this paste goes far further, and produces much better results, than sixteen pounds of liver, because it is more nourishing, and there is no waste. The food is prepared by weighing several pounds of fillet of beef, —not beef-steak, which is too stringy, nor a piece off the surloin, which is generally too fat. Fillet of horse is equally suitable with fillet of beef, and surloin of horse, being generally very lean, is nearly as good. But as no establishment kills anything like a sufficient number of horses to supply the fry with the tit-bits, the butcher must necessarily be the chief purveyor. Mutton is not suitable. All the fat being carefully scraped off, and the meat being weighed, it is pounded in a large marble mortar, and passed through a coarse sieve. The yolks of hard-boiled eggs are then added, nine eggs being allowed to each pound of meat. The eggs should be several days old, as, if new-laid, it is impossible to boil the yolk until it is mealy. This can be easily arranged by buying foreign eggs from a wholesale dealer by the box, which runs from 120 to 150 dozen, and at Howietoun generally lasts about ten days. When the yolks of egg and and meat have been thoroughly mixed in the mortar, they are passed through a fine wire sieve and kneaded into a stiff paste. This is rolled into the shape of a thick sausage, and cut and rolled into large pills, each sufficient to give one meal to five boxes. Theoretically, the weight of each pill should be checked, but in practice it is found that the eye is a sufficient guide. When the food is all prepared, it is taken into the hatching-house, and one pill placed on the edge of the fifth box in each row. One of the girls then goes round with a feeding-spoon, and, beginning at the bottom box, presses the food through the perforated zinc of the feeding-spoon, which reduces it into fine vermicelli. When the threads are about two inches long, they are shaken off into the water, and the current keeps them always in motion. The fry, having their attention attracted, seize on the moving filaments, and drag them all over the box, causing the greatest excitement, so that the fry eat quite as much out of jealousy as from hunger. I can compare it to nothing else than a pack of highly-bred hounds breaking up a fox. If the meat has been too fat, the filaments adhere, and lie in the bottom untouched. If, on the other hand, too little egg has been used, they break up into a thin soup, which very soon fouls the box. But when properly prepared, and the fry not over-fed, there is not one particle of waste.

'The feeding-spoon is made out of elm by boring a large hole out of a 1-inch plank, and making a saw-drift through which to pass the perforated zinc. The hole is tightened up with a couple of brass screws.

'Nos. 8 and 9 zinc are the proper sizes. If the holes are smaller it is impossible to pass the prepared food through, and if larger, the filaments are too thick to be easily eaten by the fry, and get broken up. Should it be imperative to feed with liver, sheep's is better than bullock's as it breaks up through the zinc into much larger particles, and, though more costly to buy, there being much less waste, is less expensive. Great care must be taken not to over-feed the fry on the prepared food, or they will stretch their stomachs to the size of the original yolk-sac, a condition which is generally followed by a suffusion of blood near the anus, and death.

'After a fortnight's feeding on the prepared food, finely ground horse-flesh is substituted. This is prepared by selecting the mash from the large chopping-machine with which the food of the older trout is prepared, pounding it in the mortar, and passing it through a very fine wire sieve. It is then fed out through No. 9 perforated zinc rolled round a circular base. This utensil we call the short feeding spoon. It is found very useful, and obviates any danger of choking, as all particles too large for the fry to swallow are retained in the cylinder, and emptied out into a pail provided for the purpose, to be mixed with the food of the yearling trout.

'Many ingenious fry-feeding machines have been designed, but as this part of fish-culture demands constant attendance, and can only be successfully undertaken where the whole time of at least one person can be devoted to the trout, I think it unnecessary to refer to them here. The best and simplest is that used at Howietoun in the experimental tanks Nos. 1, 2, 3, and 4.

'Amateur fish-culturists should, as a general rule, turn out their fry ten days before the yolk-sac is absorbed. It is a very common error to suppose fry will not feed until the absorption of the sac. Where they are deficient of vitality this may be so, but when the produce of properly selected breeders, and when the eggs have been so incubated as to induce great vitality in the embryos, and where the alevins had suitable depth of water and sufficient current, they come on the feed before the total absorption of the sac. Nature has, in fact, provided them with a large reserve of food, and, if vigorous, the hinder portion of the sac becomes separated by constriction, and drops off under ordinary circumstances, and it is only where there is an absence of vitality that the sac is totally absorbed.

'The above does not apply to *fontinalis*, nor to the ova of young trout, or of grilse, and even with the largest salmon if hatched in water of a falling temperature, whereby the period of alevinage is much prolonged, when the whole nutriment contained in the sac becomes necessary to the life of the fish.

'DESPATCHING FRY.—In the early days of trout culture fry formed the principal sales; few cared for the trouble of hatching the ova. The carriage of yearlings was far too expensive; clearing-house rates were unknown; proper preparation not understood; nothing better than a carboy had been thought of for their transport. The water in which they were conveyed required to be frequently changed; air required to be supplied either by splashing the water or by bellows, and an attendant's constant anxiety frequently supplied an illustration of the proverb, "Care killed the cat,"—especially if he solaced himself with a pipe of tobacco during an extra long spell at the bellows. In those days fry were very costly, and, as a natural sequence, they were carefully counted before despatch. Now, if there is one thing more fatal to fry than another, it is catching and counting them; the least touch removes the mucous, and fungus follows. This was very early discovered at Howietoun, and many methods were tried to mitigate the evil. They were caught up on perforated zinc, counted, and washed off into a pail; they were skimmed up with a light muslin skimmer mounted on fine wire; they were spooned up with a soup-spoon; they were shovelled up in a miniature dust-pan (used for sanding birds' cages); they were poured into milk-plates, and counted as they passed over the spout into a pail; but none of these methods were satisfactory. The miniature dust-pan was, however, far the best and quickest of the lot.

'After much consideration, and seeing that two grilles of ova were laid down to hatch in each box—which was always 7000 eggs, and frequently more than 8000,—*it was decided to sell fry by the box*, guaranteed to be not less than 5000. This left a margin of about 50 per cent., and the Fishery had the satisfaction of knowing that the fry despatched were uninjured; and although the price might appear high in comparison with that of ova, the results obtained were so satisfactory that the sale of fry has steadily increased every year.

'PREPARING FRY.—The preparation of fry is a very difficult matter. It does not do to starve them, or they lose vitality, and cannot find their own food when turned adrift in strange waters. On the other hand, if they are fully fed, they travel uncommonly badly. We have found them travel best if fed on sheep's liver for a week before they start; but it makes a foul mess on the bottom of the box, which must be carefully cleaned before pouring the fry out, as, if the water in which they are transported is the least dirty, a large proportion of them will perish. If they are fed on the prepared food up to the day of starting, their stomachs are too distended, and inflammation would be the result. If fed on pounded horse-flesh, the matter they deposit is too gross, and the water becomes fouled.

'Fry stand cold badly, and travel best in the daytime. They may be lightly fed over-night before starting, but not in the morning.

'FRY TRAVELLING TANKS.—Fry will not stand much knocking about, and if the yearling tanks are used to transport them they must be filled until the water stands above the point of the perforated zinc cone, the wave rising and falling though the zinc is cushioned, and the motion at the bottom of the tank greatly modified.

'The bottom of a tank used for transporting fry should be stiffened by cross-pieces soldered underneath, as, if it saggs at all, the fry soon get fatigued, possibly because the least spring from the bottom frightens them, and they exhaust their strength by frequent and aimless sallies through the water.

'The old tank used to carry the fry from Middlethird to Loch Leven in 1875 answers the purpose well, but the area of the bottom is so small only a few thousand fry can be contained in it. It has, however, the advantage of being light, and can be placed on a dog-cart or the box of a four-wheeled cab, and is perhaps the handiest shape for amateur work. Pieces of wood carrying iron lugs have been bolted on to the side, so that it can be carried between two sticks.

Ventilation is provided by little cylinders of perforated zinc soldered round an aperture in the lid and guarded by its handle. This in practice has been found sufficient.

'Carboys are used by some pisciculturists, and, when well filled with water, there is no jar; they keep a very equitable temperature, the thick glass being a bad conductor, besides which they are generally packed in a basket. The only objection I have to them is their weight, and the space they occupy in proportion to the number conveyed. Fry appear to travel by sea fairly well in carboys; while, in conical travelling tanks, they appear to suffer from the motion of the steamer. During the Edinburgh Exhibition of 1881, some thousand fry of *Coregoni* arrived in good condition from Russia. They were carried in a modified form of carboy packed in a square box lined with felt, air being admitted by a tube passing through a cork.

'Conical tanks have one advantage in warm weather; if a jacket of coarse sacking be laced tightly over the tank, and the lid arranged so as to admit sufficient water to escape to keep it damp, the evaporation will cool the tank so that the temperature seldom rises above 45°. The best temperature for travelling fry is, I think, above 40°, and certainly below 50°. I do not think it ever advisable to reduce the temperature of the water in the travelling tank below that at which the fry are being reared, which is usually below 50°.'

FRY: ITS APPROPRIATE USES.—The cases in which fry are preferred to Ova sown in redds are few; but in the case of district Boards where the heavy cost of yearling ponds makes it impossible as yet to rear the fry, the best course is to make redds in the upper and less accessible waters for one half the eggs incubated, and to feed the remaining half as fry until the beginning of the fourth month after hatching, and then turn them out into the most accessible of the suitable fry-rearing grounds of the district. A second and important use of fry is the stocking *new* reservoirs where the feeding streams are practicable. This has been done most successfully in the case of Lake Vyrnwy, the new reservoir of the Liverpool Corporation, to which 300,000 *S. Levenensis* fry were successfully sent from Howietoun in 1889, and which, being almost entirely new ground, yielded this last season, 1891, the most gratifying results. A third use of fry—and by no means an unprofitable one—for those who have suitable yearling ponds, but without the command of good hatching water, is the transplantation of fry, to grow to yearlings or two-year-olds for sale purposes. This we find a very satisfactory market, since once established it constitutes practically a clientele of annual customers. Fry are also much esteemed by angling associations and tenants of fishing who wish to improve their sport, but do not care to construct permanent fixtures. Although in the above two cases, yearlings would generally but not always be more satisfactory. To use fry advisedly, it is wise to consult some practical fish-culturist.

The Cost of Fry.—The great cost of fry is due to the space they require, and the loss due to netting them when required for transplantation. At Howietoun experience is in favour of fry never being placed in any trough or box larger than can be easily lifted by two men, and the fry poured out into the pitchers or carrying tanks. In practice, this entails a cost for house room so to speak double of that for incubation of the eggs, since, as compared with the number of eggs, only half the number of fry can be safely and advantageously fed in any given box.

Thus the produce of a gallon of Ova costs £2, 4s. for house rent, as compared with £1, 2s. per gallon of Ova for the period of incubation thereof. Fry further cost four times as much for labour, and consume about £1 per gallon of food.

The ledger account stands thus :—For a house capable of incubating 80 gallons of Ova and rearing the produce of 40 gallons as fry for three month after hatching, (See page 17).

Interest on £3000 at 4 per cent., ...	£120
Depreciation and renewal calculated on 80 gallons	
of Ova as 14s. per gallon (see page 18), . .	56
	£176
Of which one-half is credited to use for Ova leaving	
to debit of Fry, ...	£88
Add Wages at 80s. for 15 weeks, . .	60
Cost of rent and wages for rearing the produce	
of 40 gallons of Ova for three months after	
hatching, ...	£148

or, divided by 40 gallons, £3, 14s. per gallon of Ova—this, added to the £1, 9s. 6d., makes

£5, 3s. 6d. for the produce of one gallon of Ova reared to three-month-old fry for rent and wages alone. Food for this quantity, as extracted from Howietoun ledger, cost £42, or £1, 1s. per gallon. This, added to the lowest cost of production of Ova (page 22), viz., £6 per gallon, being the total cost of fry three-month-old and thoroughly fit to transplant, to £12, 4s. 6d., or, at the ordinary cost of production of Ova unfertilised (viz., £9 per gallon) to £15, 4s. 6d.

A gallon of Ova may be supposed to average 28,000 three-month-old fry, and it is sold in four boxes of 5000 each nominal, delivered carriage paid at £5 per box, that is, £20 per gallon Ova used to produce to fry. And the cost of delivery varies between £4, 10s. and £6, for any distance beyond local traffic rates. The weight of the water required for transplantation closely approaches one ton.

Three-month-old fry have been so successful when properly used, and are rising so rapidly in the estimation of the public, that the price will probably rise slightly in the future, as at present the margin is dangerously small. Roughly, the cost of the production of three-month-old fry may be stated at between £70 and £80 per hundred thousand nominal, and one-fifth less per hundred thousand actual count.

YEARLINGS.

Their Production, Appropriate Employment, and Cost.

THE PRODUCTION OF YEARLINGS from fry is a very simple matter. The fry are merely placed in suitable ponds and fed sparingly but continuously for all daylight hours from May until August. The food is cheap, consisting at first of raw liver, and afterwards of finely chopped horseflesh. The ponds must be suitably constructed (see page 62) and the earth renewed each season. Gulls and herons require shooting and trapping, or a dog may be trained to herd the ponds and scare off the water-fowl. One collie at Howietoun is unceasing in this watch, but in spite of dogs, traps, and guns, the gulls do take a very serious toll of the fry when first placed in the yearling ponds. One hundred thousand (nominal) three-month-old fry require four ponds, each 100 feet long by 22 wide at top by 5 feet 6 in depth, and should make from 30,000 to 40,000 good saleable yearlings and a few thousand undersized fish. The ponds must be dried for one month, and must stand full of water for at least three weeks before stocking, five weeks is better, so that the water becomes charged with animal life. The most difficult time for the fish-culturist is the few days immediately following the plantation of three-month-old fry. Numbers weaken through starvation before they know the feeding spoons, but when there is ten days supply of natural food in the pond, no fry weaken through starvation, and five boxes of fry will produce the same number of yearlings that, under less favourable circumstances, six boxes would produce.

APPROPRIATE EMPLOYMENT OF YEARLINGS.— Yearlings fit all uses, they travel better than any other size. They bear out Mr. Guy's description of them in the Howietoun Fishery Price List :--

'Yearlings are, *par excellence*, the size for general purposes. They are strong enough to find their own food, thus avoiding the principal cause of mortality among fry, namely, starvation ; they are easily carried, and stand the journey well ; they accomodate themselves with the greatest facility to new water ; and they thrive so fast in ponds that they will be found a very profitable investment.'

As a rule, yearlings are better than two-year-olds for rivers ; of course there are rivers[1] where two-year-olds may be preferred, but all two-year-olds are grown in comparatively still water. It would not pay to make a two-year-old pond on the principal of a long narrow ditch, and this is precisely the principal on which the most successful yearling ponds have been constructed.

Yearling ponds may be varied in shape to suit

[1] *E.g.*, the Thames below Henley.

circumstances, and in many cases with excellent results, *vide* Mr. Andrews' successful ponds near Guildford, where the adaptation of old water-cress beds has proved most remunerative; but under ordinary conditions the ditch form gives the largest return on the outlay, and (with the full supply of water usually given to this class of yearling ponds) most nearly resembles conditions which abound in rivers and streams in all parts of the country. Yearlings from these ponds, therefore, when transplanted to rivers, find conditions not greatly at variance to those under which they have been reared. With two-year-olds it is vastly different, they have been reared in larger ponds, where they enjoyed a wider range, and even if the supply of water to their pond was proportional to the supply usually allowed to yearling ponds, yet the greater depth and breadth of their pond so altered the amount of current as to habituate the fish to comparatively still water. Yearlings, therefore, have an advantage over two-year-olds when transplanted to small streams or rapid rivers.

Yearlings have a further advantage over two-year-olds at present rates for carriage, as, roughly speaking, 2 tons 10 cwt. are required to safely transport 1000 two-year-olds for twelve hours, while 15 cwt. is amply sufficient for the same number of yearlings on a similar journey.

Yearlings can find their own food in any new
water if it is not over-stocked.   Practically they
may be appropriately used in every case where
they are not exposed to predatory fish or birds ;
and even then it becomes a question involving a
delicate calculation, whether or not it is cheaper
to increase the number of yearlings transplanted,
so as to discount the loss, or to incur the expense
of two-year-olds.   Generally, yearlings should be
used wherever the water is unsuited to fry.

There are many hill lochs which can be stocked
by yearlings, and by nothing else, where there is
little or no feeding stream, or where the feeding
stream is too rapid or too rocky to suit fry, and
where the difficulty of access forbids the tanks
necessary to carry two-year-olds being brought
within a reasonable distance of the water.

There are many lowland ponds which can be
stocked with yearlings more satisfactorily and
more economically than by any other description
of fish—the number of yearlings being proportioned to the capacity of the water, and the risks
inevitable from birds and large fish.

There are many rivers where two-year-olds
would wander to stiller water, and where fry
would have little chance of survival—but where
yearlings would give for many years after the
stocking a good account of themselves.

And there are many ponds all over the country

which, if stocked annually with yearlings to be sold as two-years-olds the following winter, would yield a very handsome return for the cost and trouble : only these ponds would require to be annually emptied, as a few of the previous years' stocking would make short work of the new-comers.

THE COST OF YEARLINGS.—Yearlings, when grown in artificial ponds, are so under control that the cost is fairly stable, taking one year with another ; and is composed of the following items :— Cost of fry ; rent of pond, including water supply and renewals ; cost of food and attendance, including the preparation of food. There are other items before they can be despatched, such as office correspondence, cartage to station, netting and cost of preparation—for without some preparation yearlings travel indifferently.

The first of these items, viz., cost of fry. One gallon of Ova incubated and reared into four boxes of three-month-old fry costs £15, 4s. 6d. (see page 38), taking the ordinary cost of production of Ova at £9 per gallon. A little over 30 gallons of Ova incubated and reared to three-month-old fry was used last season at Howietoun, or about 120 boxes of fry ; the value of which, on the above basis, was £450, 15s. The actual cost of food is not easily ascertained from the books, as it was not kept separate from that of the two-year-olds, but it was not less than £80 for the nine months. The attendance was practically equal to the whole time, one man and one girl for the same time, say £78. The cost of rent, labour in renewals, wear and tear, cultivating the ponds by drying and re-earthing the bottoms and sides, must also be to some extent a matter of guess-work, but if taken at 10 per cent. on the structural cost of the 20 ponds, which was between £2000 and £3000, a sum of £250 per annum is obtained. Personally, I think, 10 per cent. is too little to allow for this item in most cases; but the ponds at Howietoun are very well constructed, and the annual renewing of the earth-surface of the bottoms and sides, for purposes of sanitation, is the principal item of expense. In the case of such ponds as Fishery Boards are likely to construct, from 12 per cent. to 15 per cent. should be estimated for, especially since, if the first cost is reduced, not only are more repairs entailed annually, but the capital sum on which the estimate is based is lower, thus necessitating a higher percentage. The cost of 30 gallons of Ova grown into yearlings will thus be :—

Cost of three-month-old fry, . . .	£450 15	0
Rent, repairs and renewals, etc, . . .	250 0	0
Cost of food, ...	80 0	0
Attendance, ...	78 0	0
Cost of 30 gallons of Ova = 120 boxes of fry = from 140,000 to 200,000 yearlings, .	£858 15	0

or, a gallon of Ova may be expected to produce from 4,700 to 6,700 yearlings at a cost of £28, 5s. 1d., or from £4, 5s. 10½d. to £6, 2s. 8¼d. per thousand. The latter price is nearer their average of good work than the former, as 200,000 yearlings could not be reared, in the ponds estimated on, without loss of average size ; and, in any case, there is always at least 10 per cent. of small yearlings which are unsaleable.

Writing from a District Fishery Board point of view, it would be fair to say that allowing no sum for management, risk, or profit, and working on this large scale, yearling trout may reasonably be expected to be produced at £6 per thousand, counting *all* fish whether sizeable or not. From a fish-culturist point of view, office expenses, risks, netting, preparation, despatch and cartage to station, and wear and tear of travelling tanks—no small item in itself—must be added, and the difference between that and the price is the profit on each gallon of Ova incubated and reared to yearlings. In practice, these items bring up the average cost to a little over £9 per thousand, *free on rail*. The cost of preparation alone at Howietoun, that is, 4 per cent. interest on the cost of the two substantial despatching houses, and the renewals of the wooden tanks and machines employed to save handling the fish, amounts to 13s. 4d. per thousand yearlings, and

£1, 13s. 4d. per thousand two-year-olds—but the saving in cost of railway charges for carriage is in each instance much greater. In conclusion, we are of opinion that, while Fishery Boards working on this large scale may, under favourable circumstances, raise yearlings at £6 per thousand all told, large and small, it would not pay them to place their surplus stock on the market below £10 per thousand, and if they only rear 50,000 or 60,000 yearlings annually, it is open to doubt whether these yearlings will not cost close on £10 per thousand *in the ponds* before handling at all. In the above calculations all profit in any of the stages has been neglected. If in the table (page 44) we substitute the price of 30 gallons of Ova grown into three-month-old fry for the bare calculation cost. The table will stand thus :—

The equivalent of 30 gallons of Ova, viz., 120 boxes of fry at £5 per box, .	£600
Rent, repairs and renewals,	250
Cost of food, ...	80
Attendance, ...	78
Cost of equivalent of 30 gallons of Ova in yearlings, say from 140,000 to 200,000, . .	£1008

or £33, 12s. per gallon, or from £5, 0s. 10d. to £7, 4s. per thousand. This calculation is made for the convenience of Boards and Associations, who prefer to purchase their fry instead of incurring the expense of a large hatching house. And the fact cannot be too frequently impressed that first-class incubation demands such an equality of temperature as can only be secured in this climate by substantial masonry.

TWO-YEAR-OLDS.

Their Production, Appropriate Employment, and Cost.

THE PRODUCTION OF TWO-YEAR-OLDS from yearlings consists in the careful sizing of the yearlings, and the removal of all above and below a certain standard. Should this be omitted, cannibalism is sure to arise, and, strange as it may appear, there is more risk of cannibalism in permitting a few hundreds of small yearling in the two-year-old ponds than in overlooking a few larger yearlings. The small trout are eaten by those yearlings, who may be supposed to have a larger share of original sin, and when once they commence the practice they rapidly outgrow their companions, and attain a sufficient size to prey on the ordinary stocking of the pond. These cannibals are easily known by sight; not only do they appear shorter and plumper than the others, but (with *S. Levenensis*, at least) their silvery sides assume a bright golden tint. In practice, the secret of growing evenly sized two-year-olds lies in the careful selection and segregation of the yearlings.

The ponds suitable to two-year-olds require to be deeper and wider, but not necessarily longer than those for yearlings. At Howietoun, we prefer a depth of from six to twelve feet. The supply of water is not so material in the case of two-year-olds as it is in the case of yearlings. Yearlings require a current; two-year-olds require range—and the amount of range at this stage has a marked influence on the future shape of the mature trout. Stone says in his *Domesticated Trout* [1] (a book that should be in the hands of every fish-culturist :—

'If you want to have trout short and deep, and to use an expressive Americanism, "Chunky," grow them in a deep still pond. If you want to have them long and slim, grow them in a shallow, swift current.' (*Dom. Trout,* p. 276.)

And again at p. 235 :—

'*Give them range.* If you want to grow your trout very large, you must give them range : I say if you want to grow them *very large.* Range is not necessary, by any means, to the *average* growth of trout, for they will grow to a very good size in small places, and it is also generally incompatible with trout growing as a business to give them great range ; but if you want to raise the very largest trout, you must give them the very largest range. Trout will not grow above a certain size in confinement. They will stop, or nearly stop

[1] *Domesticated Trout: How to Breed and Grow Them,* by Livingston Stone, A.M. Boston : James R. Osgood & Company, 1873.

growing, when they have reached a certain limit. Range also influences the rate of growth. Large ponds grow trout faster as a rule than small ponds. Put ten trout in a pool three feet square, and ten others in a pond ten rods square, and those in the pond will grow much faster than those in the pool on the same food.'

Stone of course writes on the American brook trout, *S. fontinalis*, which is a first cousin of the Highland *Chars*; but the above extract is equally applicable to our British trout. Stone goes on to impress the necessity of giving trout plenty of *space*, and dilates on the difference between range and space; but in this matter I do not entirely hold with him. Space means the cubic feet of measurement in each pond divided by the number of trout. Now, the real factor is composed of *surface* measurement; gallons of fresh water or cubic feet of fresh water supplied per minute; and cubic contents of the pond in feet; *and* it is further varied by the capacity for oxygen of the water—a capacity which is exceedingly variable.

The real crux is to furnish each fish with a sufficient supply of oxygen. The water may be considered as the medium from which this supply is obtained. Again, the quantity of available oxygen must be proportioned to the gross *weight* of trout, and not to the *number* of fish, and again, a slight rise of temperature will double or treble the quantity of available oxygen required. I used the term 'available oxygen,' because more or less of the oxygen in the water is neutralised, or imprisoned by carbonic acid given off by the trout themselves, and otherwise present in the pond.

So intricate is this question of space per pound of trout live weight, that nothing but practice can determine the proper stocking for two-year-olds in each case; and there must be sufficient margin left for safety during a thunderstorm on the hottest day of summer. If a pond is so stocked as to just pass through this ordeal without inconvenience, there will be no further risk from over-stocking, as the trout grow larger the weather becomes colder, and less and less oxygen per pound of trout is required.

In the production of pedigree two-year-olds for future breeders more care is required in selection, and the largest yearlings must be carefully excluded. One season we carefully selected the largest yearlings, and the result was that the two-thirds of the future breeders turned out to be male fish—a most undesirable result when the high cost of feeding the breeders on shell-fish is taken into account. Attention to pedigree and space is most important; and the rule for space in this case is—place no more fish in the pond than are necessary to consume all the food thrown in, for if a pond is understocked the fish scatter, and much of the food falls to the bottom and fouls the water far more than a much larger number of fish would do.

APPROPRIATE EMPLOYMENT OF TWO-YEAR-OLDS AS A MODE OF STOCKING.—Two-year-olds are not advisable for stocking the upper portions of rivers; they *will* seek deeper water, no matter of what breed they may be. Wherever coarse fish, such as perch and dace thrive, two-year-olds may be used with safety, but not above the limit of dace water.

For all ponds two-year-olds are better than yearlings, inasmuch as, if the feeding is at all plentiful, they will be fit for the rod by the end of the same season in which they were transplanted.

In lakes and larger pieces of water a sufficient stocking of two-year-olds is often too costly, and yearlings are used instead; but our experience is that the public is beginning to discover that for ponds, lakes, and reservoirs, two-year-olds, from the speedy sport they afford, are more advantageous than any other mode of stocking, and consequently at Howietoun no effort has been spared to provide for their accommodation.

COST OF TWO-YEAR-OLDS.—The production of two-year-olds on a large scale is of too recent development to admit of any accurate estimate of their cost on the same lines as given for Ova, fry, and yearlings. Taking yearlings at £10 per

thousand, the present price of two-year-olds, viz. :
£25 per thousand, leaves a larger margin of profit
than the price at £10 per thousand of yearlings
does ; *but* the market value of two-year-olds is
limited by value of three-year-olds at the fish-
mongers, and £25 per thousand leaves a good
margin of profit to any one who cares to fatten
two-year-olds for the market. With a little skill
and attention they may be fattened as yeld
trout for the following November, December, and
January, the process must not be commenced
too early, or the ovaries will start into activity,
and, instead of fat half-pound trout, a lot of
immature spawners will result.

The heaviest item in the cost of two-year-olds
delivered is the cost of carriage. There is no
sense in forwarding trout with too little water to
save carriage ; if the trout arrive sick they will
partly die, and the remainder will take weeks, or
perhaps months, to recover from the bad effects of
this journey. If, on the other hand, the two-year-
olds have been transplanted with a generous supply
of water, they will go ahead and thrive from the
first or second day after their transplantation. Of
course, the weight of water required varies very
much according to the temperature of the air.
During the journeys in a hard frost almost any
liberties may be taken with impunity, but in warm
weather the contrary is the case, and the warmer

the weather the *less* ice can be used without injury to the trout, as it is of importance that the sum total of the difference, of the temperatures of the ponds, the water in the carrying tanks, and of the water of their new home, should be as small as possible.

The average weight of water allowed for two-year-olds is 2 tons 10 cwt. per thousand, and the cost of carriage—including return empties—is about £6 per ton for any distance over 200 miles, or £15 per thousand, making £40 per thousand, delivered at purchaser's risk, or under 10d. per fish. For shorter journeys, or in very cold weather, the cost of carriage is less, in some cases much less. But even at the higher rate good two-year-olds are a most profitable investment.

THE CONDITIONS UNDER WHICH OVA, FRY, YEARLINGS, AND TWO-YEAR-OLDS, CAN BE MOST PROFITABLY USED.—These conditions are practically a repetition of what has been written under appropriate uses of Ova, Fry, Yearlings, and Two-Year-Olds, but it will be convenient to consider them under one head, more especially as there are circumstances under which all four modes of stocking can with advantage be used together ; for instance, Highland lochs, especially those situated not far above the sea level, generally have long stretches of sluggish water immediately above the mouths of the principal streams running into them; these stretches of water have sometimes pike and frequently perch, and are therefore unsuitable for yearlings, but as the feeding is usually more abundant in such places than elsewhere, or than in the loch, they are just the spots to choose for turning in two-year-olds, while higher up the principal burns there is often found that sequence of pool and shallow so suitable for yearlings; higher up again there are gravelly places where fry can find food and shelter, and many of the smallest burns have clear springs where Ova may be laid down, if on the point of hatching, without any redd or other preparation, and with the certainty of success. With conditions similar to the above, the greatest economy will be obtained by a nice adjustment of numbers of each of the four classes, viz., eggs, fry, yearlings, and two-year-olds.

R E D D S.

The formation of redds is a very simple matter, nevertheless much skill and local knowledge may be advantageously exercised in the choice of positions.

A careful survey of the stream or river should first be undertaken, and the amount of suitable shallows for rearing fry carefully noted. The best fry-ground are those portions of the stream or river where the bottom varies from coarse sand to small stones, and where small shallow pools and slight eddies are interspersed.

When the survey is made in winter great local knowledge is necessary to determine the amount of fry-ground available in summer. April and May are the best months to make the survey, as it is in those months the fry leave the redds and are in more need of the protection yielded by good fry-ground than at any subsequent time.

Having found a stretch of good fry-ground, the next business is to choose a situation, for the redd, a few hundred yards above. If a tiny stream or clean ditch can be found, a little coarse gravel is all that is required, but if the water has to be brought from the main stream, or from a brook or ditch subject to high floods, the redd must be cut some distance back from the water and well above the highest flood, so that even the bottom of the redd is above high flood level. The principles which govern the construction of a redd are the same as governs the construction of a hatching-box, viz., that the stream or current washes every egg and removes the carbonic acid exuded in the process of incubation. In a redd the eggs are more spread and a much slighter current is sufficient than that required in a hatching-box. The redd should

be narrow, not over two feet wide at bottom, and
any convenient length ; two square feet of bottom
per thousand eggs is the minimum space advisable,
and five square feet per thousand eggs the maxi-
mum.   A redd two feet by twenty feet will hatch
out a box of from 15,000 to 20,000 Ova safely,
but when practicable it is better to make smaller
redds and more of them.

The depth of water in a redd should be from
three to six inches, not less than three or more
than six.   The sides of a redd are best left as cut
out of the soil, and sloped as little as possible so
as to shade the redd, which should also be covered
with open hurdles or fir branches—anything in
fact which will freely admit the air and modify
the *light*.   Light is one of the worst destroyers
of trout-life, especially in the earlier stages.

The inlet of the redd may be either a four or
six inch pipe—not larger than six, or too much
water may be passed through the redd in time of
flood.   The inlet should be taken off the stream
in a wooden box guarded with perforated zinc,
to prevent small trout or sticklebacks entering
and feeding on the alevins.   The outlet should
have a fall arranged to effect the same purpose.
Mice are sometimes troublesome, but do not feed
in water over four inches.   Water-rats, however,
must be guarded against, as they are greedy
feeders below water.   It is not advisable to fence

the outlet of the redd, as the object is to permit the young fry to find their way into the stream under as nearly the same conditions as if hatched in it as possible, but a loose wall of coarse gravel and small stones forms no impediment to the alevins passing out, and affords some protection.

The quantity of Ova hatched out in redds proximate to any given fry-ground in the stream or river must be adjusted, not only to the extent of fry-ground and the probable amount of feeding, but a full allowance should be made for the amount of natural fry from the adjacent spawning-beds, and this only local knowledge can estimate. However, the artificially incubated Ova from the redds will usually, but not always, have several weeks start of the natural fry, and be able to hold their own in the struggle for sustenance.

REDDS will not take the place of stocking by yearlings or by older trout, but in most waters they will form a valuable adjunct to such stocking.

PONDS.

The general principles which govern the construction of ponds, are : shade, food, control, facility for emptying, and water supply.

Shade is the first requisite, and may be given by artificial shade ; by position, as proximation to natural shades from trees, banks, hedges, etc. ; by construction, viz., by making steep sides to the pond ; by depth of water ; by water-plants ; and by any combination of the above.

Food is the next requisite ; without shade trout won't live, without food trout won't grow. Natural food may be produced by planting suitable water-plants, and sowing amongst them *gammari* and *limnæa*, water-shrimps and water-snails. Artificial feeding requires every portion of the surface of the pond to be within easy throwing distance from the bank, so that all the trout may be fed by scattering food on the water.

Control is also very important, and means that the trout can be caught and handled whenever necessary, for sizing or other purpose.

Facility for emptying requires a drain passing below the deepest part of the pond—it is not necessary for the *first* stocking of a pond, but is very desirable before any future stocking, so that the pond may be cleansed and all fish of the former stocking removed.

Water supply governs the number of trout of a given age the pond will safely carry. Bearing the above principles in mind, it is of little consequence how the pond is formed, whether by damming, or banking, or partly or wholly by excavation—only, in the case of damming a portion of a stream, the danger from floods must be taken into account and an ample bye-wash provided. The inlet of a pond is easily screened against trout ascending—a very slight fall *over* a ledge into *shallow* water is sufficient ; trout cannot jump out of shallow water as they can out of deep ; but the outlet screen presents more difficulty and should be large enough to pass all the water in time of flood, and deep enough below the outlet to pass all the water in time of frost.

The best water-plants for ponds must vary in every locality, generally speaking, watercress-beds in connection with the inlet give excellent results.

The depth of the pond effects the gross weight of trout the pond can rear more than anything else: with shade, a pond only one foot deep will hold a few small trout, but the largest trout require even with shade, at least six feet of depth of water.

Ponds sometimes get very milky in the early winter, and as all conditions of water have important bearings on trout life, I note the observations here, so that others may consider the question. I suggest as an explanation the probability that the sudden chilling of a thick layer of water at the surface, and its consequent descent to the bottom at a time when the bottom layer of water and infraposed mud are at a temperature of at least 40° Fahr., may cause the finer particles of sediment to rise throughout the pond and produce the appearance of milkiness.

The following considerations may be of use to observers. Fresh water attains its greatest density at 39° Fahr., and the difference from the greatest density, expressed as unity, in hundred thousand parts is—

Degrees Fahr.		Density : unity less hundred thousand parts.			Degrees Fahr.
32°	=	13	12	=	46°
33°	=	10	8	=	45°
34°	=	7	5	=	44°
35°	=	5	3	=	43°
36°	=	3	2	=	42°
37°	=	1	1	=	41°
38°	=	$\frac{1}{2}$	$\frac{1}{2}$	=	40°
39°	=	0	0	=	39°

That is to say, the density of fresh water at 32° Fahr. is ·99,987, at 39° Fahr. 1·00,000, and at 46° Fahr. ·99,988.

From this it appears that the density of water somewhere between 32° Fahr. and 33° Fahr. is precisely the same as the density of water at 46° Fahr. And fluids of the same density in the same vessel must occupy the same horizontal layer. Therefore, at the beginning of winter, when the pond water is about 46° Fahr. at the bottom, the first hard frost causes the top layer to descend and mix at 32°, or thereabouts, with the bottom layer at 46°, but the difference of temperature must liberate the thermal units from the warmer water to raise the colder to the resulting temperature ; in other words, force is changed from latent

to active, and currents of varying temperatures
result, bearing the finer particles of sediment
through the whole pond and producing the milky
appearance ; and this continues morning frost after
morning frost, until the bottom layer attains the
temperature of the greatest density, viz., 39° Fahr.
after which no action is apparent, or, if any, the
ponds appear clearer after a frosty night than before.

In this suggestion I have assumed that no per-
ceptible effect takes place except after the first
sharp frosts—that is, only when the surface layer
is so *rapidly* cooled that no great action takes
place in passing 39° Fahr. The temperature of
the greatest density would occur in the case of a
gradual cooling, when the bottom water would be
reduced to this temperature without any distri-
bution, and the colder but lighter water could not
pass through to stir the sediment.

I also assume that the top layer of water would
descend in semi-vertical currents *through* the in-
termediate layers.

The position of feeding trout in lakes, sometimes in mid-water and sometimes at the surface,
has possibly much to do with temperature currents,
more so than with the average temperature of the
layers of water through which, I suppose, these
currents to pass. With the variation that sea-
water attains its greatest density at a temperature

below 32° Fahr., this suggestion may also explain the movement of herring shoals, and ultimately furnish some guide to the depth nets should be sunk for successful fishing. It may be advisable to experiment to ascertain the laws and conditions under which a rapidly cooled layer of salt water could pass through the immediate infraposed layers.

THE CULTIVATION OF WATER.

A piece of water can be cultivated so as to yield results which can only be compared to a market garden or to a vineyard.

Probably the larger lakes will always in this country be more valuable for sporting purposes than as food producing waters ; as trout preserves, than as trout farms. Their cultivation as preserves will consist of careful and systematic netting, to remove all trout destroyers, such as pike, perch, and above all, old large-headed, lanky-bodied, barren trout ; of the introduction of fresh blood when desirable ; and of stocking with such discretion that the average size of the trout caught with the rod be not below the desired minimum.

The introduction of water-plants and of natural. foods,—for even the May fly may be successfully transplanted while in the caddis state, and in some cases the planting of timber, to shelter exposed parts of the lake margins, also form part of the cultivation of lakes as trout preserves.

I may now add a few words on their cultivation as trout farms : here the object is to market the greatest weight of pink-fleshed trout, and size is of slight importance, in fact, in many places half-pound and pound trout command as good a price per pound as larger fish.

In cultivating for the market, it will, in most instances, pay to net out all trout over two pound weight, to reduce the loss from cannibalism, and as the netting for the market requires to be constant, the shotts for the net must be kept clear from water plants, which require to be introduced with more caution than when the lake is used for sport, and only netted in winter and spring to keep down vermin.

In cultivation for the market, *gammarus* and *Limnæa peregra* are valuable flesh-colourers, the former especially giving a deep pink to the flesh of the trout. I have found swan mussels [1] useful in increasing *gammarus*, but whether from the shelter their dead shells afford, or whether they are eaten by *gammarus*, who is a great scavenger, I cannot say.

As yet we are only on border-land of water cultivation ; the subject is well worthy of study and promises to repay manyfold any labour that may be expended upon it.

[1] Anodonta cygnea.

The cultivation of rivers consists in securing a natural flow at all times throughout the river; but unfortunately this is precisely what cannot be done. If compensation water be honestly passed down, weirs and mill dams pierced with passes, pollution stopped and weed cutting regulated, much towards improving the river fishing for Salmonidæ will have been accomplished. Many persons think that a portion of a stream has only to be fenced off at either end and it can be readily cultivated and safely stocked; but the difficulty of making a dam water-tight and a screen trout-proof is great. Any little fall may be utilised to prevent trout ascending, but as yet it passeth the wit of man to prevent their descending out of a stream. The difficulty is due to flood, leaves, and frost; it may be mitigated, but cannot be entirely obviated. The safest plan is to form a long level overflow the height of the outlet screen, so that, when the latter clogs up, the depth of the water on the overflow may be less than one-inch; trout will not readily pass a knife-edge overflow under these circumstances.

TRANSPORT OF TROUT.

The colder the weather the better for transporting yearlings and two-year-olds, the cooler the weather the better for fry; in other words, in the hardest frost two-year-olds may be sent to any distance in Scotland, England, or Wales, not only without loss on the journey, but without any loss after transplantation ; we always prefer to break the ice for two-year-olds. With yearlings it is different, to this extent that yearlings travel better by day than by night in very cold weather, when, and only when, their is a long wait at any point *en route,* in all other cases they, like two-year-olds, travel best at night. . . . Fry cannot stand frost at all, nor do they care for ice in the travelling tank, we therefore like to send them so as to arrive early in the morning in April or up to the middle of May, when the weather is cool in distinction to frosty . . . with a little attention to the above points trout of all sizes can be safely transported.

In conclusion, we recommend that all large pieces of water should be examined and reported on before being stocked, so that not only the best strain of trout may be selected, but that the mode of stocking may be varied to suite the circumstances of each particular case. I have appended a draft report to show the questions requiring to be dealt with, and the considerations which arise in the case of a typical Lowland Lake.

E.

APPENDIX.

—◆—

4th January 1892.

DEAR SIR,—I have yours of 29th ultimo, and enclose a Preliminary Report on the Lake of . . . Time has not permitted of my personally visiting the Lake, but this is of the less importance, since a full report necessitates the investigation and tabulation of the Mollusca and Crustacea, more especially the Daphniadæ Copepoda, and Ostracoda, one species of which, *Candona Candida*, forms a large portion of the natural feeding at Howietoun.

This investigation can only be carried on by boat and in summer, and by the frequency of the various forms of Mollusca and Crustacea, the gross weight of trout the lake can carry is controlled. I must therefore omit the Invertebrate Fauna, and the flora on which the fauna largely depends, from this Preliminary Report, and deal with the mode of stocking and question of cost in a general manner, leaving any final recommendations till the completion of the report.—Yours sincerely,

J. R. G. MAITLAND.

PRELIMINARY REPORT

ON THE

LAKE OF . . . AS AN ANGLING LOCH FOR TROUT.

THE season being unsuitable for a proper investigation into the food-bearing capacity of the Lake of . . ., I have prepared this interim report from my general personal knowledge of the district. I feel no doubt that the Lake of . . . is capable of being converted into a first-class sporting trout loch; but at present I am unable to state approximate cost. I append, however, certain calculations which will serve as a guide to the Meeting.

GENERAL CONSIDERATIONS.

Pike and perch are plentiful, and the pike at least will require thinning. Yearling Loch Leven trout will hold their own against perch—smaller burn trout will not—it is merely a question of the size of the "yearling."

PIKE.

The modes of thinning down pike are so well-known that I need not enter into the matter here, further than to remind the Meeting that reducing the number of pike is more easily accomplished before spawning, which is from March to May in this district, and that netting should be commenced in February. If it is desired I will be happy to enter fully into the subject in the final report.

Perch destroy small yearlings; large yearlings of four inches and upwards are comparatively safe. The size of the yearling (other things equal), depends on the size of the parent egg. This again depends on the kind, size, age, and nutrition of the spawner. The best *S. Levenensis* eggs at Howietoun count 25,000 to the gallon, while at Loch Leven the average is about 35,000. Small hill burn trout eggs count 50,000 to the gallon, but by selection and careful treatment, the size can be raised to 40,000 eggs to the gallon. From these figures it is evident how far careful selection has removed *S. Levenensis* at Howietoun, from Loch Leven trout. If, however, a smaller breed of trout is desired—to hold their own against perch—the young trout must be retained two season's instead of eight months in the rearing ponds, this multiplies the cost per hundred or per thousand two-and-a-half times.

NATURAL FEEDING IN LOCH.

The trout at present rise very badly, from which I suspect the bottom feeding is good. If this be so (and it is only by a careful examination in summer that amount of bottom feeding can be estimated), the trout will rise freely so soon as the stock of trout is increased slightly out of proportion to the bottom feeding. If the stock is still further increased, both the rate of growth and the average weight of the trout will diminish. The prudent rate for stocking a piece of water in this part of Scotland, under 300 feet above the sea, is 150 trout (yearling) per acre of less than 12 feet depth. When the water is over 12 feet, the bottom feeding rapidly decreases in quantity. When the feeding is very good the number of yearling per acre of water of 12 feet or less, may rise to, but never exceeds 200. But when the water is more than 300 feet above sea level, or when the bottom feeding is poor, the number falls to 100 or even fewer yearlings per acre, the depth of which does not exceed 12 feet. If the stock is too light there is little sport, the fish grow larger and rapidly, but do not rise freely. If the water is over stocked, there is again little sport—although the fish rise freely and are numerous, since they remain so small as not to be worth fishing.

MODE OF STOCKING.

There are four modes of stocking, viz. :—

By Redds.

When there are many small gravelly streams—affording altogether a large acreage of feeding streams suitable for fry—this is the most economical mode of stocking, but is not suited to the Lake of . . ., the spawning streams being naturally insufficient for the acreage of the lake.

By Fry.

This for the above reason is inadvisable.

By Yearlings.

Yearlings being placed directly into the loch, and being reasonable in first cost are recommended for the Lake of . . . . The only objection is that some loss must occur in after years from cannibalism.

By-two-year-olds.

This is the usual mode of stocking, where old trout or coarse fish already exist. It is gaining in favour every season, and for moderate sized pieces of water where

nearly every fish comes to the rod, the cost of stocking
is fairly balanced by the dead value of the trout caught;
but I am afraid in so large a piece of water as the Lake
of . . ., the cost of annual stocking with two-year-
olds would be out of proportion to the result. It might,
however, be the most convenient for the first year.

I have not the acreage of the Lake of . . ., but I
suppose there is about 500 acres of fishable water, and
about 200 acres either too shallow or too deep. On the
above supposition, at a moderate stocking of 100 yearlings to the acre, 50,000 yearlings would be required.
These trout, three years after, should yield an average of
between 9 lbs. and 10 lbs. per acre (which is of 12 feet or
less in depth). 5000 lbs. annually is the gross weight of
trout that should reasonably be expected to the rod. That
is,with 50,000 yearlings as a stocking in 500 acres of fishing
water (there being no pike supposed for purposes of this
calculation), it would be reasonable to expect 20,000 $\frac{1}{4}$ lb.
trout caught in the second season, 10,000 $\frac{1}{2}$ lb. trout in the
third season, and 5000 1 lb. trout in the fourth season,
leaving 15,000 trout for casualties and to grow to larger
fish. The question is, what annual stock of yearlings is
equivalent to a single stocking of 50,000? The number
varies in different waters, the cause being due to cannibalism. If all the trout could be caught or removed in the end
of the fourth season, the second stocking with 50,000 yearlings would be in 1896, which gives the annual average
required of 12,500, but owing to the cannibalistic pro-
pensities of trout, the annual number required is at
least **20,000** (in place of 12,500). And this I believe
to be the lowest annual stocking of yearlings the loch will
require. Should the feeding turn out as good as I anti-
cipate, a much larger number—up to 40,000 yearlings
per annum—might be used with economy and success;

but the number, whatever it is, should be adjusted so as to secure a *minimum* average of 1 lb. per trout caught with the rod.

NUMBER REQUIRED—TWO-YEAR-OLDS USED.

The number required for a single stocking of two-year-olds (under the suppositions above) does not materially differ from the number of yearlings, viz. : 50,000 for 500 acres of 12 feet depth and under—as the loss at Howietoun between year-olds and two-year-olds for the year is under 5 per cent. But when the case of annual stocking is considered, $\frac{1}{4}$ of 50,000 or 12,500 per annum is nearly equivalent, as although single stocking with two-year-olds requires to be triennial, and it would give 16,500 approximately, yet the 15,000 allowed for casualties with yearlings is not required, and thus restores the balance. However, if the two-year-olds are retained as $\frac{1}{4}$ lb. trout, that is, when caught the same season they are placed in the loch, some addition must be made to the annual stocking.

COST OF STOCKING—OVA IN REDDS.

The redds being mere ditches communicating with the natural stream, are not in themselves expensive. In favourable situations two men can cut a redd sufficient for 20,000 eggs in two days, and the first cost of 20,000 eggs is about £7, 10s. But the two burns running in to the Lake of . . . are not capable of rearing sufficient fry to keep up a fair stock of trout. Therefore, though two or three redds might be made advantageously, they would only provide a small supplementary stocking.

COST OF STOCK—FRY.

Fry cost about £75 delivered per 100,000, but have no advantage over a corresponding amount of ova in redds— 100,000 three-month-old fry is practically the produce of

150,000 eggs. The principal mortality being in the first month after commencing feeding.

COST OF STOCKING—YEARLINGS.

Yearlings (in this particular case) would cost about £11 per 1000 delivered. This, for an annual stocking of 20,000 yearlings would entail an expenditure of £220.

COST OF STOCKING—TWO-YEAR-OLDS.

Two-year-olds would (in this particular case) cost delivered, something under £30 per thousand, which for 15,000 two-year-olds would entail about £450 per annum.

COST OF STOCKING—BY MEANS OF REARING PONDS.

I have no doubt that this is the proper course to adopt, and may be considered under three sub-heads, viz. :—(*a*) Cost of construction ; (*b*) Cost of fry; (*c*) Cost of management, food, risk, etc.

(*a*) COST OF CONSTRUCTION.

The ponds for rearing yearlings require to be about 100 feet long, 22 to 25 feet wide, and *must* be 6 feet in the centre, with an iron drain-pipe and wooden valve to draw off the pond which *must* be dry for at least three weeks every year. The quantity of water required is one 6-inch pipe with a pressure of three diameters or 18 inches. Experience has shown that the inlets and outlets should be built, and that brick-work is preferable to masonry. The cost of each pond is about £120, exclusive of water supply and water course. The capacity of each pond is from 8000 to 10,000 good yearlings and about 2000 small yearlings annually.

(*b*) COST OF FRY.

The equivalent in fry is difficult to calculate, as so much depends on the occidentation of the pond, the quality of the water, and the skill of the attendant. At

Howietoun we use six boxes, or 30,000 fry (nominal) to stock each pond. The loss is almost entirely due to cannibalism, which in a rearing pond is more prevalent than in an open stream. After the yearlings are caught and sized there is no further loss. Six boxes of fry would cost £30.

(c) COST OF MANAGEMENT, FOOD, ETC.

Each pond would require between stocking in May or June and emptying in April, 15 cwt. of horse flesh and about 1 cwt. of liver on first stocking. Horse flesh can be bought in Glasgow,

15 cwt. would cost about—	.	£3	0	0
And 1 cwt. liver about—	.	1	0	0
In all		£4	0	0

Say £4 per pond per annum. There is still to add the proportion of the attendant's time—about 2 hours per day per pond—say £12 per annum. Until the attendant was trained, it would not be reasonable to expect more than 7000 yearlings per pond. Thus three ponds would be required to ensure 20,000 yearlings.

COST OF STOCKING BY MEANS OF REARING PONDS—
CAPITULATION.

Three ponds at £120, £360; add inlet and outlet works and inlet pipe, say £40, in all £400; and allow interest and up-keep at 3 per cent., repairs 9½ per cent. = 12½ per cent. on £400,

12½ per cent. on £400,	.	.	£50	0	0
Food, three ponds at £4,	.		12	0	0
Attendance, 6 hours per day,	.		36	0	0
Cost of ponds per annum,	.		£98	0	0
And 18 boxes of fry at £5,	.		90	0	0
			£188	0	0

F

as against £220 for 22,000 yearlings—the difference is perhaps hardly worth the risk—but with a good man 30,000 yearlings might be reasonably produced, and the surplus either sold to reduce the cost, or if the bottom feeding turns out as I expect, then 30,000 yearlings would by no means be an excessive stocking.

I have not suggested a hatching house, because all experience is against small hatcheries, they are in the end more expensive than buying from larger establishments, and the first structural cost is very heavy, if good incubation is to be obtained. The principal hatching-house at Howietoun cost, fitted, over £3000. Nor have I suggested two-year-old ponds, as the depth and size, from 10 to 12 feet deep, and for 12,500 two-year-olds an area of that depth of half an acre, makes these ponds very costly, especially as if the two-year-olds are to thrive the pond must be deepest and drained from the centre; if any other form of pond is used, it must either be larger in extent, or much more lightly stocked.

BREED OF TROUT.

All the Scotch trout with two exceptions (Islay and Orkney), pass from one form to another so as to render, after a few generations, identification very difficult. I therefore lay more stress on the *selection* of breeders than on the original strain. But for the Lake of . . ., the Loch Leven strain is, I believe, the best. Those which have been bred at Howietoun since 1874 from the best yearlings—the produce of the largest breeders—are more stable in their characteristics than trout bred chiefly from Loch Leven, where promiscuous breeding is tending to slightly lower the average weight, while a little skill would in a few years very materially increase it.

In closing this preliminary report, I have to impress the importance of first killing down the pike before they spawn in March. Should rearing ponds be determined on, they should commence early in summer. I do not recommend a heavy stocking with yearlings before the pike are reduced. In future years the trout will to some extent keep down the young pike, and the net should be able to do the rest. All large trout netted should be killed until the stock of trout is good.

If rearing ponds are made, it will be necessary to employ an engineer, and most of the work should be done by days' labour; of course the supply and waste-water courses can be contracted for, but the cutting the ponds, banking, inlet, outlet, and iron piping are not capable of being contracted for.—Yours truly,

J. R. G. MAITLAND.